Beginner's Guide To Building Your Own Battery Pack

GRAYSON H. SISSNEY

Table of Contents

Introduction

A few hundred years ago, the thought of electrical items that could make our lives easier was nothing more than a mere dream, something that seemed to be completely impossible. Today, however, we rely on technology for almost everything we do. Most people sleep with their smartphones right next to them, and the screen of their smartphone is often the first thing they see in the morning. Your alarm clock, television, gaming console, car – these are all objects that rely on electricity to power them and make them function. Even though all of these objects require electricity to function, a variety of methods can be utilized to power these devices and deliver the required electricity to the devices. While the electricity provided in buildings make it easy to plug in a charger and refill your smartphone's battery, as well as to power up your television, dishwasher, and other electrical appliances, access to electrical outlets are usually limited when traveling, and especially when going on a hike or somewhere remote.

In cases where electrical outlets are not available, even during travels, other options are available to help power your devices. Battery packs, power banks, portable chargers – these are all popular options that people opt for in the modern day to carry power with them wherever they are. Unfortunately, one particular problem that people often face when it comes to these products that will help them keep their electrical devices powered during their travels is that battery packs can be quite costly. A 48V lithium rechargeable battery pack for an electric bike, for example, costs well over $200 at most retailers. A battery pack for a drone or RC plane can reach well over $50 – sometimes even more than $100.

The idea of making your own DIY lithium battery pack can be daunting at first. It sounds very complicated, and surely you will need some skills if you wish to succeed at this task but, rest assured, the process of creating your own custom battery pack is most likely easier than you think it is. While there is some skill to the process of making your own battery pack, and you will require a set of different tools, I am here to guide you through the basic steps that you need to understand in order to create a fully functioning DIY lithium battery pack on your own.

The biggest benefit that you can obtain from making your own battery pack will obviously be the fact that the process will save you a significant amount of money in most cases. Even though you need some special tools and equipment to get the job done, most people find that they are still able to get away with half the money they would have spent should they have opted to buy a pre-built battery pack. Another beneficial factor that I would also like to mention about making your own battery pack is the fact that you can customize the pack according to your requirements. In many cases, a battery pack is needed in a specific shape, in order to accompany a custom designed piece of equipment, for example. If you find yourself in such a situation, you already know how hard it can be to find a pre-built battery pack with the structure and shape you need it in, while also delivering adequate power output at the same time.

In this guide, I am going to provide you with an overview of how you can build your own custom DIY lithium battery pack at home. I'll point out the skills that you need in order to succeed with the task, provide a complete overview of what tools and equipment you need in order to be able to build your own battery pack, and I'll give you an overview of some popular types of battery packs that are often created. Additionally, I'll also cover some additional topics of interest that you may find helpful in your pursuit of building your very first battery pack.

Before I start with the official part of my guide, I do want to point out that, even though I have now mentioned that building your very first battery pack is most likely easier than you think it is, you should expect some difficulties and perhaps even disappointment during your first try. This is completely normal – by reading my guide, it means you have little to no experience in building these devices but would like to get your hands dirty in building your first battery pack. For you to succeed, you have to allow yourself to make mistakes during the process. If you do something wrong, do not fear. Take your time, be patient, follow my guidance – skills are not something we can learn in a mere few seconds, but rather acquired through practice and patience.

Chapter One – Why Build Your Own Lithium Battery Pack?

You might be wondering why would you want to build your own lithium battery pack when you can simply pop into a local store or visit Amazon, or any other online shop for that matter, and buy one that has been pre-built. In reality, it would be much more convenient to go out and simply buy a lithium battery pack and then install it on the particular device you wish to use this power storage unit in. This would save you quite a lot of time; I am not going to argue with that factor. There are, however, a couple of reasons why you may want to consider rather building your own battery pack instead of opting for one that has already been built and comes ready-to-install. Before I start sharing the process of building a battery pack with you, I would like to provide you with a quick overview of some important reasons why you should consider following through on this guide and completing all the steps, ultimately building your own custom battery pack.

DIY Lithium Battery Packs Saves You Money

I have already mentioned this before, but want to stress this factor – if you opt to build your own battery pack, then you are saving money in the process. In most cases, you will be able to save at least 50% when you compare the price you pay for the equipment and parts needed to make your own battery case to the price of a pre-built battery case. This is the simplest reason why you should the process of building a battery pack at home in mind. The fact that you are interested in my guide most likely means you have done some initial research into the different battery packs on the market, which means you already know just how expensive these energy storage devices can get. Even though yours may not end up looking as fancy as some of those that are currently being sold on the market, you'll be able to save money when you build the device yourself. For many people, especially those who are shopping for a lithium battery pack on a tight budget, this would be the greatest advantage to be obtained from a DIY approach.

Portability And Convenience

The very purpose of a battery pack clearly describes this particular benefit I am mentioning. You might be thinking that portability and convenience is not really a true reason why you would want to build your own battery pack, considering the fact that the majority of pre-built battery packs on the market today can provide you with these two features out-of-the-box. That is both true and false, however. Yes, buying a lithium rechargeable battery pack from Amazon will give you a device that is both convenient to use and portable, but thinks about the fact that these devices have not been tailored toward your particular requirements. If you build your own battery pack, it means you will be planning a structure according to what you need and what the device will be used for. This means that your DIY project can offer a higher level of portability and convenience, as compared to the in-store options when it comes to tailoring the battery pack to fit better in your scenario.

Reduced Weight Over Other Options

This particular reason why you may want to build your own battery pack will depend on what you need the device for. If you are going to be building a battery pack and equip it to a UPS, for example, then you will achieve the benefit of building a device that has a lower total weight as compared to using some of the other options that can be fitted in such technology. Valve-regulated lead-acid batteries are the most common type of battery used in UPS systems and many other systems that rely on batteries, but can add a considerable amount of weight; thus reducing the mobility of the hardware or device. When creating a lithium battery pack for these purposes, weight can be reduced significantly, which provides for a more mobile device.

Faster Charging Times

While I am on the topic of comparing the features of lithium batteries to some of the other options, such as lead-acid batteries, another important factor to take into account is charging time. You obviously want your battery pack to reach a full charge as fast as possible, so you can grab it and go. Lithium batteries are known to provide a much faster charging rate than many other battery options that you can opt for, including lead-acid batteries.

Education And Skill Acquirement

In addition to the reasons I have mentioned above, I also want to point out the fact that making your own DIY lithium battery pack ultimately provides you with added education, as well as allows you to acquire new skills. Since you have shown an interest in this guide, there is a very good chance that you lack at least some of the skills that are needed to create a battery pack. This means that you are hereby presented with an excellent opportunity to learn more about how these battery packs work and how they are put together. After following this guide from start to finish, the skills will become very valuable to you. Even though I will only be touching some of the basics that are involved with the process of building a battery pack at home, the education and skills can be used to experiment with other projects, where your creativity comes into play.

An Inventory Of Battery Tools

Apart from the fact that choosing to build your own battery pack instead of buying one from a local store being able to save you money while working on the project at hand, there are longer-term benefits that are also to be taken into account. In particular, I want to focus on the fact that there are numerous tools that you will need in order to build your own lithium battery pack – I'll cover the particular tools that you will need for this project later on in my guide. For now, just consider the fact that you do need a variety of tools to build this device. Many people find that, even though they need to buy some tools for their first build, they still save money when compared to the price they would have had to pay should you have opted for a battery pack on the shelf at your local store. The majority of the tools you will need for the project will be reusable. This means you are faced with an opportunity to build an inventory of battery-building tools. The next time you decide that you want to build a DIY lithium battery pack, there will be no need to go out and buy all of these tools. Since you have already bought them, you might simply need to buy some materials, and you are ready to get started with your second battery pack.

Chapter Two – What You Can Do With A DIY Lithium Battery Pack

Next up, I want to discuss the possibilities of a custom-built lithium battery pack quickly, before we set out on the journey to building your very first pack. I have already told you that there are a number of different reasons why you might consider rather building your own battery pack, instead of buying one off the shelf. Now, I want us to focus on what you can do with such a battery pack. Focus on the idea of building one yourself. This allows you to customize the shape and appearance in order to satisfy the needs of a particular project you may be working on, or whatever purpose the battery pack will ultimately serve.

The truth is, there are simply too many different opportunities for an individual who chooses to build their own battery pack – the possibilities of what you can do with such a device is truly countless. For the purpose of this guide, I would like to provide you with some possibilities – just to give you an idea. Take these ideas, perhaps give them a try, make one of them your first project, but don't let they limit your opportunities for invention. Use your creativity and make your own lithium battery pack for any device or appliance that could be compatible with such an item.

One of the most popular reasons why people want to build their own lithium battery packs would be to install the pack on an electric bike. The recent trend in electric bikes, primarily caused due to the fact that they were only made popular in the last few years, have contributed significantly to the fact that electric bike battery packs are becoming so popular. While battery packs specially designed to fit electric bikes can be purchased on a number of online shopping portals, as well as in many physical stores in your local area, they will not provide you with the same versatility of what you can achieve by building your own battery pack.

There are other uses for building a lithium battery pack, of course. Many people make their own lithium battery pack to have a portable power supply with them wherever they go. This way, they can be sure that when their smartphone's battery dies, then they have a backup power storage device to recharge their smartphone's battery. If you get a little more creative, you could use lithium battery packs for many other purposes – power your laptop on-the-go, create an emergency power backup device for your valuables that could lose data when the power suddenly goes down, set up alarm systems and always have power with you no matter where you go.

Apart from the actual usability of a DIY lithium battery pack, there are other things that I also want to point out in this chapter. While the most important topic to discuss when looking at what you can use a lithium battery pack for would obviously be what the pack can power, you also need to consider the fact that you can customize these packs toward your specific requirements. Need a battery pack that is round to fit into a cylinder? Rather prefer a square-shaped battery pack? You are in control.

In addition to being able to customize the shape and, of course, the power of your battery pack, you are also free to place any type of casing on the battery pack. Create a unique metal mold for your battery pack, use a tight-fitting plastic - the possibilities are really endless.

Chapter Three – What Skills Do You Need To Make Your Own Lithium Battery Pack?

Now that you know why building your own lithium battery pack is so beneficial and you know what you can do with such a tool, you are probably wondering whether or not you would actually be able to build your own. It sounds quite complicated, and the thought of dropping into a local store to pick up a battery pack that has been pre-built for you sounds more convenient. Luckily, if you practice some patience and put your mind toward completing the project – from start to finish – then things may turn out to be somewhat easier than you might have thought.

You don't have to be exceptionally talented to be able to build a battery pack. All of the skills that you need to build your first battery pack can be acquired, but be warned; it may take you a few tries to get everything right. For this reason, I highly suggest that you go easy on yourself, especially if this is the first time you have been interested in exploring the construction of a lithium battery pack. Allow yourself to make mistakes, and when you do, learn from these mistakes, instead of allowing them to be the downfalls that cause you to quit your project.

There's really not much to the process of building a battery pack. You will need to do some soldering, however. This is a particular area where some people may get stuck a little. Soldering is usually an easy task, but some people do tend to find it more difficult to get used to how the process works. Some background in general electronics would also be very useful. Education on the basics of electronics will help you better understand how everything should ideally come together, how each part should be perfectly fitted for seamless operation, and what you need to avoid in order to prevent possible safety hazard alarms from going off.

Acquiring the required skills for building a DIY lithium battery pack does not have to be a tiring or even a stressful process. If you have absolutely no soldering skills, then you might want to consider taking up a short course in your local area. While this may cost you additional money, think of the course as an investment. You will gain valuable skills in the process that will not only come in handy during this particular project but will also be incredibly useful in the future, should you decide to build more battery packs or perhaps decide to start another type of project.

If money is a little tight, then you may want to consider looking up some videos on the internet. YouTube is an excellent place to start. Just do a search for “how to solder," and you’ll be presented with thousands of results. Look for the top-rated videos that include videos such as “starting” and “beginners." These videos can also be very helpful and guide you through the process of soldering the right way.

A general understanding of electronics and, specifically, how batteries and power work, can also be obtained in a number of different methods. You could sign up for a short course on electronics at a local college. There are also many online colleges that offer correspondence courses. Once again, if money seems to be a problem, then consider looking up some educational videos on YouTube and perhaps some educational websites. You could always see if a platform like Udemy has courses that may help you achieve a better understanding of how batteries work and fit together in custom projects.

Chapter Four – The Planning Phase

Finally, we have reached "the good stuff." From this point on, we will start to discuss the actual process that needs to be followed should you wish to build your own DIY lithium battery pack. One of the most important steps that should ideally be taken with any type of project, including the construction of a DIY battery case, would be to include a planning phase in the project. The planning phase will allow you to ensure everything during the actual construction process goes well. During the planning phase of your project, you have to decide on a few vital factors that will ultimately have an impact on the final product that you will be producing. When building a lithium battery pack, there are a couple of objectives and elements that you need to cover in your planning phase – this will help you avoid any potential complications and safety hazards, while also ensuring that you build a battery pack that works exactly as you need it to work.

I would like to guide you through the planning phase of your very first battery pack to ensure everything goes as easy as possible for you. I will cover all of the most important sections that should be included in your planning phase to provide for a successful project that actually works – if you don't do things right from the start, you could end up with a safety hazard instead of a portable power device.

I highly recommend you get out a pen and paper for the planning phase, or at least open up a notepad on your laptop, tablet or mobile phone. As you follow through this chapter of my guide, I will frequently ask you to note down certain things – such as an idea, a purpose, parts that you may need, and more. Make sure you follow my instructions, as this will ensure you do not run into any issues down the line.

Keep Your Budget In Mind

When your creativity kicks in and you start to feel inspired, it can be quite easy to get carried away. Unfortunately, when you get too carried away, then you may come to a point where you are unable to afford to construct the lithium battery pack you are planning to produce. For this reason, I highly recommend that you decide on a budget you can spend on the entire project before you get started. When you look at the budget you are setting aside for the project; I recommend taking inventory of any tools you currently own that might be useful during the construction of your DIY lithium battery pack. If you already have some tools that will be needed to create your battery pack, then you obviously do not need to buy them again. If you are not sure what particular tools will be needed, refer to the chapter titled "Tools And Material Required For Your DIY Lithium Battery Pack" a little later on in my guide.

Since this is most likely your first project of this kind, there is a good chance that you may need to buy a couple of tools. First decide on a total budget for the entire project, including the tools you need (which you will be able to use in future projects, remember). Once you have a projection of how much you can spend on your project, then the next step is to determine what particular tools you will need to buy to create your battery pack and to identify an approximately budget that needs to be set aside for these tools. What is left in your budget can be appointed to the material of your battery pack?

If you have an estimated budget in mind for the materials that can be used to create your own battery pack, then it will be easier to decide on a size, power, closure and other features of your battery pack. Work as closely as possible within this budget during the entire planning phase, and you'll see that things are much easier when you start to implement the tasks needed to make your battery pack, as you won't be faced with unexpected fees, which could suddenly force you to postpone your project.

Identifying The Purpose Of Your Project

If you want to build something custom, then the very first thing you need to do in order to make sure your project becomes a success is to identify your project's purpose. In this case, you need to decide why you want to build your own lithium battery pack. Be specific, as the purpose of your battery pack will ultimately decide how you will plan out the layout of the project, how much power you will require, what shape you should ideally adopt, and how you should cover your battery pack. I have covered some potential uses of lithium battery packs in a previous chapter of my guide. Even if you simply wish to experiment a little with lithium batteries and building a battery pack, you should still come up with a purpose for the battery pack you are about to construct. Perhaps you wish to create a battery pack that will allow you to keep your mobile phone charged when you are not home, or maybe you wish to become part of the new electric bike trend.

Planning Your Project

Once you have decided the purpose of your project, then you are ready to start with the rest of the planning phase. During the next part of the planning phase, you need to go into more detail as to what you are going to build and how you would like to build it. Be as specific as possible, as the data you note down during this step will be used in later steps to help make the process of constructing your battery pack much easier. You should break your project into smaller parts, and then do some planning on each of these parts. The more specific you are, the easier your project will be to execute once your planning is done. You'll also be able to compile a more accurate list of the tools and material you will need in order to succeed at your project.

Start by deciding the perfect size for your project. Now that you know what you are going to build a DIY lithium battery pack for, it should not be difficult to determine the ideal size for the project. In addition to the sizing of the project, you should also roughly decide on what type of shape you would like the battery pack to resemble. The sizing and shape should be ideal for the battery pack. If you are going to use the battery pack on an electric bike, make sure that you plan an appropriate size and shape to fit a convenient location on the electric bike properly.

You also need to determine how powerful your battery pack needs to be in order to power the desired device. Different devices require different power levels. If you decide to create a portable power bank for your mobile phones, for example, you might decide to construct a 6000mAh power storage unit with a 3.7 voltage feature. You need to make sure that the size of the battery pack you will be creating will be adequate for the power needed by your device(s). The size will not always affect the shape as much, but may in some cases. When building a portable power bank for your mobile devices, you obviously do not want to carry a brick-shaped tool with you. A cylinder-shaped or perhaps a flat-shape may work better in such a case.

There are some other factors that you also need to account for in your plan. How would you like your battery pack to look at the end? Some people will prefer to use a simple plastic wrap that can be molded to perfectly fit the layout of their battery pack, while other people will rather prefer to make a more attractive tool they can carry around with them. In such a case, you need to consider whether you will require a custom mold, or will rather opt for a bendable material that you can fit around your battery pack. Additional features can also be added, such as special cables and ports, perhaps a metal casing, stickers, and printing – these are only a couple of examples. Account for all of these added features that you wish your custom DIY lithium battery pack to possess, as this will help make the process of deciding what materials you need, and acquiring those materials, a more streamlined process.

Chapter Five – Understanding The Different Types Of Lithium Batteries

One more chapter before we jump in and start building a lithium battery pack. I thought it would be a good idea to do a quick overview of the different lithium battery types that are available on the market. When you do go out to find a couple of lithium batteries to use in your project, the different types of these batteries currently available on the shelves can be very confusing. The reason why I think it is important to cover the different types of Lithium batteries on the market is quite simple – each type has its own pros and cons associated with it, as well as specific features that make it ideal for particular types of projects. Additionally, some types of lithium batteries are more affordable than others, some are safer than others, and some types of these batteries provide you with a more reliable source of power.

Cobalt Oxide Lithium Batteries

The most common type of lithium batteries is cobalt oxide lithium batteries. If you are reading this guide on your smartphone, tablet or laptop device, then you are most likely already using a type of cobalt oxide lithium battery. These are the common type of lithium batteries that are used in the majority of portable electronic devices, including mobile phones, tablet devices, laptops, digital cameras, MP3 players and more. They feature a low discharge rate, as well as the highest level of energy density when compared to the other common types of lithium batteries that are currently available commercially. It should be noted, however, that cobalt oxide lithium batteries are a dangerous type of battery to utilize, especially when the battery suffers damage. Furthermore, the material utilized in these batteries are a relatively scarce resource, which means obtaining cobalt oxide lithium batteries for your battery pack might turn out to be a somewhat expensive option. Still, this is the particular battery type that is often recommended.

Titanate Lithium Batteries

Titanate lithium batteries are not as commonly used as cobalt oxide lithium batteries, but do offer some particular advantages that should be noted. The inherent voltage features of these batteries are 2.4 volts, and the energy density is also low. Additionally, it should be noted that titanate lithium batteries can operate in environments where low temperatures are present. These batteries are known to operate well in environments up to -40 degrees Celsius.

Iron Phosphate Lithium Batteries

Iron phosphate lithium batteries are ideal for projects that will involve hotter temperatures, as these batteries are known to work well in environments with higher temperatures. They do not carry the same risks as the other options when looking at lithium battery types when it comes to performance under higher temperatures. These batteries feature a lower volumetric capacity when compared to cobalt oxide lithium batteries and most of the other lithium battery types. They are most often utilized in equipment used by medical professionals, as well as in some power tools. A particularly beneficial trait of iron phosphate lithium batteries is the fact that these batteries are known to last longer than most other types of lithium batteries.

Manganese Oxide Lithium Batteries

A type of lithium battery that utilizes manganese oxide. This is a more affordable option when looking at lithium battery types, and also provides a longer lifespan than many of the other lithium battery types I have mentioned here. Some companies utilize manganese oxide lithium batteries in mobile phones, laptops, and other portable devices. These batteries are also commonly used in Hybrid cars. The discharge rates of manganese oxide lithium batteries are high, and their energy density is also very low.

Nickel Manganese Cobalt Oxide Lithium Batteries

The final type of lithium battery that is currently commercially available to the general public is the nickel manganese cobalt oxide lithium type. These offer an impressive lifespan, and they are also very safe to use in a variety of applications. Since cobalt is a relatively scarce material, these batteries can be somewhat pricey. These batteries can be utilized in devices that operate at higher temperatures. Nickel manganese cobalt oxide lithium batteries are quite popular amongst companies that manufacture power sources for electric bikes, as well as amongst hobbyists who make their own battery packs for electric bikes. This is also a particular type of lithium battery that is sometimes used to power electric trains. In addition to these uses, nickel manganese cobalt oxide lithium batteries are sometimes used in power tools as well.

Other Types Of Lithium Batteries

Above, I have mentioned only five of the most common types of lithium batteries that you will find on the market today. All of these battery types are currently commercially available, which means you will not have any trouble finding the particular option you decide to opt for in your project. These are, however, not the only type of lithium batteries out there. Those that I left out are either not particularly popular, does not serve the purposes needed for the construction of your own battery pack or are currently not available to the general public. I also want to note that, at the moment, new types of lithium batteries are also going through research and testing. These will advance the way we utilize power storage sources in the modern world.

Chapter Six – Tools And Material Required For Your DIY Lithium Battery Pack

Before you start a project like this one, you need to gather some essential tools and material. This will ensure you do not need to run out to a local hardware store to gather extra material while you are working on your battery pack – and will also avoid you having to wait until the next morning should you notice you do not have a specific piece of hardware or material needed to complete your project. In the previous step, you did some initial planning for your project. This planning will now come into play and will be useful during the process of gathering the materials and tools you will need to complete your battery pack. If you haven't done the planning process until the end, then I highly suggest you move two chapter back and go over the details I have shared in the particular part.

In the previous chapter, I have provided a brief overview of the most common lithium battery types that you will come across when you set out to buy a set of lithium batteries, which will be used as the power source in your battery pack. I need you to take a close look at these different types of lithium batteries and to decide which type will be best suited for the project you wish to work on. In addition to deciding on the type of lithium battery that you will be using in your project, you also need to decide on cell size. The most common type of cell used in modern-day projects, especially those that are custom-built at home, are 18650 cells.

There are many different sized cells out there that you can opt for. The reason why 18650 cells are popular is that they offer a good balance between power, size, and price. For this reason, the majority of people who would like to build their own battery pack, especially those who are new to this type of construction, choose to opt for 18650 cells. When you first see one of these cells, you will notice it often closely resembles the popular AA battery that is used in thousands of applications. While the cell looks like a battery, it should be noted that this is a "cell," and that a collection of these "cells" will be required to make up the complete "battery" or "battery kit" that we are building.

When deciding on the particular type of lithium battery that will be used, as well as the type of cell that will be used, you should take your project ideally into consideration. Go back to the plans that you compiled earlier on here. Consider what the final project requires – do you need a fast-charge device, or will an overnight charge be okay. If you need to be able to charge up and go quickly, then you should opt for a type of lithium battery that will give you a faster charging time. If you are okay with waiting for the charge to complete, then you may opt for another type of lithium battery that takes longer to charge, but is perhaps able to give you better power output, better capacity or improved overall performance.

The batteries, or cells, are obviously not the only materials that will be needed to complete your project but do play a big part in allowing you to define a particular shape and size of your project. Take the previous markings into account when making a final decision.

In addition to the actual lithium batteries required to make your battery pack, you will require additional tools and materials to complete the job. It is relatively difficult to provide you with an accurate list of materials that you will require to make your battery pack, since different projects may require different materials. For example, if you wish to make an attractive battery pack, you will require more tools and material than when opting for a simple battery pack that is wrapped with some moldable plastic.

Tools Required To Construct A Battery Pack

Let's start by looking at the particular tools you will require to make your first battery pack. I am going to share the most useful tools that are needed to get the job done. In some cases, additional tools may be required, but it really depends on the specific type of battery pack you wish to make, as well as any extra features you wish to install in your project. As I have mentioned before, the tools that you need may add some additional expenses to the first battery pack you construct. Some of the tools I am about to mention may be somewhat costly. There is, however, the benefit that you will not be required to purchase the majority of these tools again in the future. Thus, instead of thinking of them as extra expenses in your project, rather consider them as an investment. Chances are, you will most likely be wanting to experiment with building more battery packs in the future. Once you master the skill and "art" behind the construction of battery packs, you will notice how useful they are, and will be wanting to make more of them for your other devices.

Battery Spot Welder

The most expensive piece of equipment that you will need to build your battery pack successfully would be a spot welder. A lot of people start out with their first battery pack by using a general welder – and tend to find out that this is one of the biggest mistakes they could make when it comes to building one of these devices. The reason behind this lies within the fact that a standard welder tends to have larger jaws and points, and they can produce a lot of heat at their center point. When this much heat is placed on a lithium cell, a chemical reaction can occur, which then causes the cell to lose some of its capacity. The damage dealt by this heating can also lead to poor performance by the cell and might cause the cell to become unusable faster as compared to using the right type of welder.

Another important factor that I would like to note here is that you should ideally look for a spot welder that has been specifically made for batteries. The reason for this is that these spot welders feature a different type of welding point, which usually contain electrodes. It might be a little difficult to find a battery spot welder, but with some research, you should be able to find one. There might be some local shops in your area that sells these or at least a source that could find a supplier that sells this type of spot welder. If it seems like you are unable to find any local source for such a device, then I suggest you give eBay and Aliexpress a try. Both of these websites can be very helpful in aiding you in your quest to finding the right tools for your battery pack.

When you do find a source for a battery spot welder, you should note that the market currently holds two primary types of battery spot welders. The first is a more affordable one that is constructed to allow the hobbyist to produce their own projects. The second type is the more professional battery spot welders. Hobby battery spot welders are obviously much more affordable than professional ones. The majority of battery spot welders that are designed with the hobbyist in mind costs under $250. Professional battery spot welders tend to cost much, much more – often more than $1,000.

I suggest you opt for a hobbyist battery spot welder, mostly due to the fact that they are more affordable than professional options available on the market. At the same time, I need you to be very careful when buying a cheaper option. Don't buy a device that seems too cheap – while this might make you feel like you are saving money; you could end up with a piece of crap that does not do what it is supposed to do. When buying from a website like Aliexpress, be sure to look at the reviews left by past buyers. This will help you get a general understanding of the quality rating associated with a particular battery spot welder you might be interested in. I also recommend opting for a website that provides a buyer protection program, as well as to buy from a seller that provides a guarantee with your purchase. This way, you will be able to return the product for a refund should it not provide you with the expected performance and features.

If you are unable to afford to buy a battery spot welder right now, then it might be a good idea to reach out to some friends. If one of your friends have delved into the world of battery packs or a similar area before, there could be a chance that they currently have a battery spot welder. Perhaps you could borrow the spot welder from them to use in your own project. You may also reach out to some hardware stores in your local area. In some locations, there might be stores that will offer you a rental option – this way; you will usually be requested to pay a deposit, as well as a daily rental fee for borrowing the item. This could be a much more affordable option when compared to outright buying a new battery spot welder.

In addition to ensuring you have a battery spot welder, also make sure you have some soldering solder, as this is a tool that is required during the soldering process.

Other Tools You Will Need

The other tools that will be required to build your lithium battery pack are much more affordable than the battery spot welder. Here is a list of some additional tools that you should ensure you have in your inventory before you start working on your battery pack:

- A hot glue gun, along with an extra refill stick to ensure you do not run out while working on your project.
- You will need a digital voltmeter during the process to perform certain tests.
- A heat gun will be needed if you decide to use a shrink tube or moldable plastic that needs to be heated in order to shrink around your battery pack. If you do not have access to a heat gun, then using a hair dryer on a warm setting may also be an option to consider.
- You should also ensure you have a pair of scissors, as you will need to cut some objects, such as the non-static tape that will be required during the construction of your project.

Materials Required To Construct A Battery Pack

Once you have collected all the required tools for your project, the next step is to collect the materials that you will need for the actual battery pack. I have already discussed the fact that the batteries, or cells, serve as the primary materials – these will be the power source in your device. There are, obviously, many other materials that you need to obtain in order to make your device work effectively. Here is a list of the materials that you will require to build your first battery pack:

- Nickel strips
- Silicone wire
- Electrical connectors
- Non-static tape

These are the basic material that you will need for the construction of a DIY lithium battery pack. You will also need some type of cover for your project. Not everyone does apply a cover to their battery pack, but it is still recommended to keep everything in place and to avoid potential hazards. A cover for your battery pack will also make the final product look much better. There are different options available when it comes to covering your battery pack, although most people who are building battery packs as a hobbyist prefer a shrink wrap with a large diameter. There are also shrink tape options with large diameters that can be used alternatively.

Another very important part that you will need to equip your DIY lithium battery pack with is a battery management system. I will discuss this particular part in more detail later on in my guide, but, for now, be sure to pick up a high-quality battery management system. This part will play a crucial role in your project. The battery management system is ultimately responsible for the charging of the battery pack, as well as the discharging of the battery pack. In addition to these features, some battery management systems are also equipped with additional features, such as the ability to provide a reading regarding the current charge level to a user interface. There are newer types of battery management system that provides the ability to read the battery's current state-of-health, current voltage, and more.

I also want to provide a quick note regarding the nickel strips mentioned above. There are many different types of nickel strips on the market at the moment, which can make it very easy to choose the wrong type. It is essential that you look for PURE nickel strips. The two primary types of nickel strips that you will get to choose from will usually include pure nickel strips, as well as steel strips that have been plated with nickel. The plated nickel strip will obviously seem like the more convenient choice here. These nickel-plated strips are more affordable than pure nickel strips, but there are advantages to rather opting for a pure nickel option. With a pure nickel strip, your batteries will perform better. They will have a prolonged lifespan, and overall heat buildup will be reduced.

The thickness of the nickel plates that you buy will also matter. A thicker nickel plate generally provides a better current, but, at the same time, it is important to note that some battery spot welders will not be able to do the job when the nickel plate is too thick. Thus, consider the potential of the battery spot welder that you will be using before you choose the thickness of the nickel plates you will be using in this project. You should ideally opt for nickel plates that are approximately 0.1mm in thickness. These are usually easy to utilize with most battery spot welders. If the current is too weak, then you can always stack multiple layers of these nickel plates on top of each other.

Chapter Seven – Safety Concerns That Should Be Taken Into Account

There are some safety concerns that I want to address before proceeding with the steps required to make a battery pack. In this section, I would like to address two areas where safety concerns need to be considered in order to avoid possible complications and, of course, to be prepared for that small chance that hazardous errors might occur. I would like to focus on safety concerns during the actual construction process of the lithium battery pack, and also take a quick look at what you should keep in mind after the construction process has been finished.

First of all, I would like to note that lithium batteries are usually considered safe. The chance that lithium batteries will cause problems or hazards is very small, but since there is always that small chance, you should keep the safety factors in mind and exert some effort to minimize the risk of errors and problems developing. The primary concern regarding the safety of these batteries lies within the quality of the battery cell. You can choose to play it safe and opt for certified lithium batteries, or go the cheap way and decide to buy a couple of non-certified lithium batteries. Non-certified options are more likely to hold manufacturing faults than certified options, which is why it is always recommended to rather utilize a certified option.

I also recommend wearing rubber gloves during the construction process of your battery pack. You will be working with electrical items and also hot items (think about the temperature of the soldering iron). If you do not keep your hands protected, there is always that small chance that the electrical wires may cause a shock, or that you could burn your hand with the soldering iron. Additionally, some objects may be sharp enough to cut you. In addition to wearing rubber gloves, safety goggles would also be among my preferred list of safety measures to take while working on your battery pack.

During the actual process of building the battery pack, it is important that you consider the fact that too much heat can cause problems with the molecular structure of the battery cells. When this happens, it can cause a decrease in the battery cell's capacity, as well as lead to other potential hazards occurring – including the battery catching fire or exploding. This is why I highly recommend using a specific spot welder that has been built for the purpose of batteries.

When the final product is in use, it is important to notice that any type of external stress can cause an increased heat in the battery cells and, eventually, lead to the development of potential complications – as listed above. Ultra-fast charging solutions are one particular area that should be avoided with certain types of lithium batteries, as this can cause harm to some of the chemicals located in the battery cells, as well as lead to possible hazardous complications.

Chapter Eight – Getting Started With Your DIY Lithium Battery Pack

It is now time to get started building your very first DIY lithium battery pack. By now, you should have a good understanding of everything there is to know for a beginner about building these battery packs. It is important to obtain this education before you start building your first battery pack – this information will give you the ability to ensure you choose the right material and tools to build your pack, as well as ensure you know what to do and what not to do; thus allowing you to build a working battery pack that won't pose any hazards.

The next step now is to plan out the layout of the battery pack – specifically how the battery cells will be stacked. You first need to consider the power requirements of your battery pack, which will help you identify how many battery cells you will need to equip in your battery pack. A popular option is to aim for a 36 volts battery pack, which is usually adequate for an electric bike, for example. For this to be possible, you will need to use a 10-series layout that has a 3.6 volts rating per battery cell.

The series is the horizontal layout of the battery pack in most cases. In addition to planning the number of series needed for the battery pack, you also need to consider the parallel setting for the layout of your battery pack. Let's say you wish to achieve a capacity for your battery pack that reaches 11,000 mAh; then you will have to plan it out so that your parallel setting provides you with this feature. If each of your battery cells has a capacity of 2,800 mAh, which is quite a popular capacity for lithium batteries, then you will have to include four battery cells per parallel setting of your battery pack.

Let me put this in simpler terms – you want to achieve a capacity of 11,000 mAH and a voltage of 36 volts. Thus, you will create a battery pack that contains 10 columns (3.6V x 10 = 36V), and four rows (2,800 mAh x 4 = 11,200 mAh).

The battery cells in your battery pack need to be connected with the nickel strips that I discussed previously. You should do a plan of the layout to ensure you position the battery cells in the correct way. Positive and negative sides of the batteries should be connected in the right manner to avoid potential complications and errors. Parallel configurations, which refers to the capacity of the battery pack, should be matched along all four (the number of battery cells in each parallel set of your battery pack) cells. Along the horizontal line or the series of the battery pack, you need to alternate between positive and negative. The first row is positive at the top; the second row is negative at the top, the third row is positive at the top, etc.

Tip: Before proceeding to the next step, it is a good idea to use the digital voltmeter to test the voltage of each battery cell you will be

using in your battery pack. Make sure that the voltage range of the battery cells is in a similar range. Note that some factories may ship their lithium batteries in a somewhat discharged state as to improve the lifespan of the product, but the voltage reading should still give you a general idea whether or not the batteries all have a similar value when it comes to their current voltage level.

Chapter Nine – Welding And Assembly

Time for action! During this step, I need you to wear your gloves and your safety goggles and to ensure that you are careful. You are working with hot tools, and there are hazards involved in the process. Practice safety measures – rather be safe than sorry. By now, you should have already tested the voltage of the batteries and confirmed that they are all similar, and you should have planned out the layout and structure of the entire battery pack. These steps are important and will make the assembly and welding process of the battery pack much easier for you.

Start by cutting the nickel strip into smaller strips. Measure the distance of your parallel groups so that it fits from the start of the first battery cell to the end of the last battery cell. You will start out with the parallel groups, and then move on to connect these to the each other; thus forming the final product.

You will now start the welding process. This step requires as much precision as possible, so be careful. It would be a good idea to have some spare battery cells that you could use should something go wrong. Extra nickel strips will also be beneficial should you do something wrong during the welding process.

Your parallel groups of battery cells need to be held together during the welding process. For this, you will have to utilize a grid or, alternative, make use of the hot glue gun I mentioned as a required tool for this job. If you do not have a grid or another type of device to hold the battery cells together while you weld, then glue them together using the hot glue gun.

Once glued together or place together on a grid, you should be ready to start the welding process. Place a nickel strip on the top of the row, and then start to weld. Two arms will be placed on the first battery. Initiate the welding process, then move on to the second battery. Be sure to do a test on how strong the weld is – if you find that you can easily pull the battery away from the nickel strip, then the current of the welding machine is too weak, so turn it up a little. If you feel that too much heat is being dispersed or that signs of burning are present, then dial the current down. Continue welding the strip to each battery until you reach the last one in the first parallel setting. Turn over and connect the bottom sides of the battery cells with another nickel strip.

You should continue doing this for the other parallel groups as well. If you have decided to do a 10-series battery pack, then you will repeat this process ten times, each time welding together a set of three or four battery cells (or how many battery cells you have chosen for the parallel setting of your battery pack). I do recommend performing more than one welding operation on each side of each battery. This will ensure the welding is stronger and provide a better connection. Contact and current-flow will also be better this way.

Once you have completed the welding of all parallel groups in your battery pack, the next step is to connect them together; thus making the series of the battery pack. For this step, it is crucial that the parallel groups are alternated between positive ends being at the top and negative ends being at the top. If the first parallel group has its positive ends facing an upward direction, then the next parallel group should have their negative ends facing upward.

There are two ways to connect the series of the battery pack. The first is to create a completely linear connection, which will cause the battery pack to take on a rectangular shape with no holes or bubbles on the sides. The other option is to utilize the open space that will be created when the batteries are placed against each other. In such a case you would have each parallel group of battery cells fit into the holes created by the previous parallel group's battery cells. By creating a perfectly rectangular battery pack, you'll be saving some space on the sides, but the final product will be longer than the other option. Ultimately, you should decide on which one of these particular two options you wish to opt for based on the purpose of your DIY lithium battery pack, and where it will be fitted to serve its purpose.

Start with the first two parallel groups. Place nickel strips that connect the first battery cell in the first parallel group to the first battery cell in the second parallel group. You could also describe this as connecting #1 series, #1 parallel to #2 series, #1 parallel.

Now, you should not connect the second parallel group to the third parallel group here, but instead, skip over to connecting number three to number four. Thus, this could be described as connecting #3 series, #1 parallel to #4 series, #1 parallel, and then #3 series, #2 parallel to #4 series, #2 parallel, etc. Again, you won't be connecting parallel group number for to number five here, but rather five to number six. Seven to number eight. Eight to number nine.

Once this step has been completed, it is time to do the bottom part of your DIY lithium battery pack. Turn over the entire battery pack, and now it is time for the final connections of the nickel strips. You now need to make those connections you did not make at the top. Thus, you'll be skipping the very first parallel group. Parallel-group two and three will be connected, then four and five, six and seven, eight to nine.

By reaching this part of this chapter, it means you have completed the most essential part of your battery pack. Congratulations, you have built your very first battery pack. As you will notice, however, you cannot actually use the pack yet. There are still some additional steps that need to be implemented before you can connect the battery pack to a device and power it up.

Chapter Ten – Installing A Battery Management System

Now that you have welded all of your battery cells together and you have the basic DIY lithium battery pack right in front of you, the next step you need to take is to install a battery management system, often abbreviated as a BMS. The battery management system is essential to your battery pack as this device allows for charging and the discharge of your battery pack. The purpose of a battery management system is also to ensure power is adequately distributed to all of the different battery cells that make up the battery pack, as well as to ensure that power is adequately cut as parallel groups start to fill with power. This, essentially, helps to avoid any of the battery cells in the battery pack to become significantly discharged, which could lead to potential problems in the future.

Adding a battery management system to your battery pack can be somewhat complicated. You will first have to ensure that the battery management system you choose to install is compatible with the configuration of your battery pack. There are many different options out there – make sure the battery management system you choose can provide the appropriate functionality to all of the series you included in your battery pack, or you may achieve an imbalance in the charge rate of your battery cells.

You should look at the battery management system's primary chip for numbers - #1 should be coupled with the first parallel group of your battery pack. The second wire should be coupled with the second parallel group of your battery pack. Go on until all parallel groups have been connected to your battery management system. It should be noted that it is usually recommended to offset the connection of the battery management system's wire a little on the nickel strip in order to avoid placing too much heat directly on the battery cells. The heat can cause problems with the battery capacity and may even lead to imbalances in charge distribution – this can lead to issues, including potential hazards like fires and an explosion of the affected battery cell. The discharging and charging cables also need to be connected to the appropriate terminals on your battery pack to ensure both charging and discharging can occur without any problems. These two cables are usually somewhat thicker than the other cables that need to be connected from the battery management system to the battery pack.

Most battery management system boards come with an included wiring diagram. Be sure to refer to this wiring diagram in order to ensure you connect the right wires to the right locations on your battery pack. You should also add any additional material and tools that you have chosen to include in your DIY lithium battery pack. There is a variety of items that can be added to a battery pack – each will be connected in a unique way. Ensure you closely follow all of the included instructions in order to avoid potential hazards and to make the right connections.

Chapter Eleven – Covering Your DIY Lithium Battery Pack

You should now have your DIY lithium battery pack set up and ready to go. Before you start using the battery pack, however, it is highly recommended to cover it using some type of material. While there are many fancy ways of covering your new DIY lithium battery packs, such as by using a custom-made metal plating or plastic coating, there is no need to spend extra money on such features. A great way to cover your lithium battery pack is to use a heat shrink plastic or tape. There are a variety of options available that can be used for covering your lithium battery pack. An affordable option would be to use a plastic wrap – these wraps are placed around your battery pack, and then shrinks to perfectly fit around your battery pack once the heat has been applied to the wrapping material.

For maximum protection and to ensure your battery pack is kept secure, covering the battery pack with some craft foam before applying the plastic shrink wrap is a good idea. The craft foam should be folded around the lithium battery pack, and then you can keep everything together by applying some electrical tape on top of the craft foam.

The last step is to use the shrink wrap to finalize the covering of your DIY lithium battery pack. Many people find that obtaining heat shrink material is relatively difficult. You will most likely not be able to find a local supplier in your area that stocks this type of material, especially since you are looking for a particular type of shrink wrap that has a large diameter. For this reason, I suggest you look at websites like Aliexpress and eBay. Once again, when you do place an order at any of these websites, be sure to check the current ratings of the buyer you are planning to purchase the product from and see if any guarantees or buyer protection programs are in place to keep your money protected and to keep you safe during the process.

Once you obtain the shrink wrap, you should wrap it around the battery pack. Ideally, this step should be implemented after you have applied the craft foam. This will not only provide for an added level of protection but will also help to make the battery pack more stable.

After wrapping the battery pack with the shrink wrap, you will need to use a heating tool to shrink the material. As I have explained earlier on in my guide, you should ideally opt for a heat gun. These tools have been specifically designed for a purpose like this one. Unfortunately, buying a heat gun can be expensive and may even turn out to be a waste of your money. If you do not have a heat gun and you do not know anyone who can perhaps borrow you a heat gun, then you should not go out and buy one. Instead, simply switch on your hairdryer and use this device instead of an actual heat gun. Make sure you set your hairdryer on a relatively high heat setting. Apply the hairdryer to the shrink wrap, but take care not to cause overheating. The shrink wrap will respond to the heat expelled by the hairdryer, and start to shrink toward the size and shape of the battery pack.

When you apply the shrink wrap to your battery pack, it is vital for you to ensure that the wires that will ultimately be used as the power connectors to provide power to a device should be left outside, but without causing excessive amounts of open space to be present in the wrapping of the battery pack. This is important as once you have applied the shrink wrap, you will have to cut it open should you wish to change anything within your lithium battery pack.

Some people will prefer to use their battery pack bare without the use of craft foam, shrink wrap or any other type of covering. While the actual function of the battery pack is what you should consider most important, there are some benefits to covering your battery pack with craft foam and shrink wrap. One of the very first benefits that you should take into account is that, when the instructions I have recommended here is followed, it means your final product will be water resistant. Note that I am not saying your battery pack will be water-proof, but rather water resistant – there is a difference, but making your battery pack at least water resistant can be extremely beneficial.

Vibrations, overheating – many factors can cause stress on the battery cells included in a battery case. When applying shrink wrap to the battery pack, it means you are providing a better foundation or structure for your battery pack. The improved stability of the structure of your battery pack will provide for a reduction in these vibrations, and also act to prevent some of the other factors that can cause stress on the battery cells. In turn, this means that your battery pack will maintain an equal distribution of power among all battery cells more appropriately, and also means the capacity of your battery pack will be preserved for a longer period of time. The state-of-health associated with the battery cells you used in your battery pack will also be maintained better when you ensure appropriate measures are taken to prevent excessive stressors from damaging the battery cells.

Conclusion

The final chapter of my guide. By reaching this far, it means you have (hopefully) followed through the entire guide, took in the information and guidance I have shared with you, and you have equipped yourself with the knowledge required to create your very own lithium battery pack. I have covered a decent amount of topics in this guide to ensure you are able to adopt the information and education shared here to help you build a custom battery pack for different purposes. Even though I have only outlined a relatively small number of use-cases in this guide, such as the construction of a battery pack for the purpose of keeping your smartphone charged and, of course, for the purpose of powering your electric bike, you should not limit your options to these scenarios only. There are many more uses for custom-built battery packs, as I have outlined early on in my guide.

I would like to both thank you for following my guide and congratulate you for reaching the final chapter. I hope you have gained the skills and the knowledge that is required to create your own battery pack. If you have not yet viewed my guide in detail, I suggest you do so now. Building your own lithium battery kit is not only about saving money on the purchase of a pre-built battery kit, but also about learning new skills – these skills can be perfect, which will allow you to build more sophisticated and better battery packs in the future.

It is now time for you to start experimenting with the many opportunities that lie in front of you. Custom-built battery packs are extremely convenient and can be built according to your exact requirements. You decide how powerful the battery pack is, what shape it will take on, and all other details that will make it suitable for the particular project you are building the pack for.

Made in the USA
Middletown, DE
30 April 2023